Alan McKirdy has written many popular books and book chapters on geology and related topics and has helped to promote the study of environmental geology in Scotland. His other books with Birlinn include *Set in Stone: The Geology and Landscapes of Scotland* and *Land of Mountain and Flood,* which was nominated for the Saltire Research Book of the Year prize. Before his retirement, he was Head of Knowledge and Information Management at Scottish Natural Heritage. Alan is now a freelance writer and has given many talks on Scottish geology and landscapes at book festivals and other events across the country.

Argyll and the Islands

LANDSCAPES IN STONE

Alan McKirdy

BIRLINN

For Hugh Andrew

First published in Great Britain in 2017 by
Birlinn Ltd
West Newington House
10 Newington Road
Edinburgh
EH9 1QS

www.birlinn.co.uk

ISBN: 978 1 78027 466 9

British Library Cataloguing-in-Publication Data
A catalogue record for this book is available
on request from the British Library

Designed and typeset by Mark Blackadder

FRONTISPIECE.
Isle of Jura from the summit of Sgarbh Breac, Islay.

Printed and bound in Britain by Latimer Trend, Plymouth

Contents

Introduction

Argyll and the islands that lie off the west coast of the Kintyre Peninsula have an ancient human history. This area was the seat of the earliest kings of Scotland, and Kilmartin Glen (between Oban and Lochgilphead) includes prehistoric archaeological sites of international importance. Some of the finest burial cairns are found here, and in 2012 an excavation site provided proof that people have occupied this land from the Bronze Age onwards.

But the rocks beneath the surface tell a story of an even more ancient world that stretches back billions of years. The timescale that these rocks represent is almost unimaginably long. From the very earliest history of Planet Earth to more recent times, geological events are incompletely, and sometimes enigmatically, recorded in the rocks of Argyll and the islands.

Much of the ancient bedrock of this area was created from a once-towering mountain chain that was formed when continents collided. Granites were formed deep in the Earth's crust as a result of the white heat of collision. After erosion by ice, wind and water over millions of years, some are now exposed at the surface. Volcanoes were active at various times and they too have left an indelible print on the landscape. Coal swamps briefly covered the landscape only to be succeeded by desert sands. The final chapter of this epic tale was the coming of the glaciers that shaped and scoured the landscape into the familiar mountains and glens we recognise today.

So the bedrocks of Argyll and the islands tell a long and complex geological story. In this book we look at the sequence of events that created this remarkable ancient landscape.

Opposite. Kyles of Bute

Argyll and the Islands through time

Period of geological time	Millions of years ago	Scotland's global position	Environments and events in Argyll and the Islands
Anthropocene	Last 10,000	57° N	During this time, *Homo sapiens* (people!) appeared on the scene and started to modify the landscape by clearing the forests and building structures. These were initially simple shelters and burial chambers, but in more recent times these man-made interventions became more complex and industrialised.
Quaternary	Started 2 million years ago	Present position of 57° N	• 12,600 to 11,500 years ago – this time was marked by a final advance of the ice, sweeping across the land from the higher ground to the east. • 14,700 to 12,600 years ago – the climate at this time was similar to that of today. The glaciers melted away causing a rise in sea level. • 29,000 to 14,700 years ago – a thick sheet of ice covered Scotland during what is regarded as the final glacial advance. • Before 29,000 years ago – there were many advances and retreats of the ice, separated by warmer interludes, known as inter-glacial periods.
Neogene	2–24	55° N	Subtropical conditions prevailed across the country, which then gave way to warm, temperate conditions. At the end of this period, things gradually got colder as the Ice Age approached.
Palaeogene	24–65	50° N	Volcanoes erupted in adjacent areas, such as Mull and Arran. Thin ribbons of magma (known as dykes) were intruded into the rocks of Argyll.
Cretaceous	65–142	40° N	Warm shallow seas covered most of Scotland.
Jurassic	142–205	35° N	Dinosaurs roamed the place we now recognise as the Isle of Skye. But no rocks of this age are preserved in Argyll and the islands to tell us whether or not dinosaurs also existed here.
Triassic	205–248	30° N	Seasonal rivers flowed across the area.

Period of geological time	Millions of years ago	Scotland's global position	Environments and events in Argyll and the Islands
Permian	248–290	20° N	Small patches of desert sandstones preserved on the west coast of Kintyre indicate there were widespread arid conditions across the area at this time.
Carboniferous	290–354	On the Equator	Scotland lay close to the Equator at this time and tropical rainforests were widespread across the country. A small coalfield near Machrihanish provides evidence of this long-disappeared environment. There is also evidence of erupting volcanoes at this time.
Devonian	354–417	10° S	The extreme southern tip of the Kintyre Peninsula is built from sandstones and conglomerates of this age. These deposits were transported by rivers and accumulated in seasonal lakes.
Silurian	417–443	15° S	Volcanic rocks were generated as a result of the closure of the ancient Iapetus Ocean. The rocks erupted at the surface forming what we recognise today as the Lorne Plateau and they also formed granites deep in the Earth's crust.
Ordovician	443–495	20° S	Colliding continents formed a mountain chain of Himalayan proportions. The rocks that resulted from this process are known as the Dalradian.
Cambrian	495–545	30° S	The Iapetus Ocean started to close.
Proterozoic	545–2,500	Close to South Pole	• 600 million years ago – an ancient continent was split by the formation of a new ocean – the Iapetus Ocean. Sands, muds and limestones accumulated in this major ocean. There is also extensive evidence of volcanic eruptions that became layered together with the sedimentary rocks. This pile of sediments and lavas was later cooked and squashed as continents collided. • Around 1800 million years ago – the oldest rocks in the area were formed from volcanic rocks that we now recognise as the gneisses of the Rhinns of Islay.
Archaean	Prior to 2,500	Unknown	No rocks of this age are found in the area.

Geological map of Argyll and the islands. The representation of the area's geology in map form is complex. The oldest rocks are to be found at the Rhinns of Islay. The majority of the rocks elsewhere across the area are known as Dalradian – rocks cooked and squashed when continents collided at the time the Iapetus Ocean closed. Thick layers of sands, muds, limestones and lavas were caught up in this collision and were bent and buckled into folds that are many kilometres across. Granites were also created in the white heat of the collision, and some magma was erupted onto the surface, creating the Lorne Plateau around Oban. A thin band of sandstones and conglomerates run around the tip of the Mull of Kintyre from Campbeltown southwards. A small coalfield is located at Machrihanish, indicative of a larger belt of tropical rainforest that existed 320 million years ago when this part of Scotland sat astride the Equator. A small remnant of Permian desert sandstones is to be found on the west coast of the Kintyre Peninsula. Thin vertical ribbons of volcanic rock, which originated from the adjacent Mull volcano, are also prominent features of the landscape. The final event to have shaped this part of the world was a pervasive glaciation. It finds no expression on this map, but it has had a huge effect on making the landscape look the way it does today.

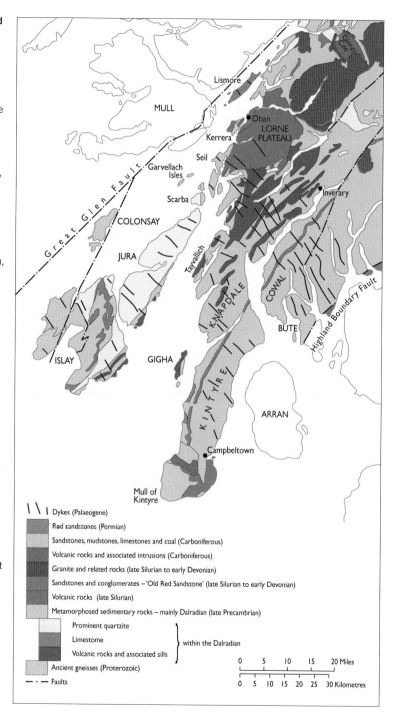

Dykes (Palaeogene)

Red sandstones (Permian)

Sandstones, mudstones, limestones and coal (Carboniferous)

Volcanic rocks and associated intrusions (Carboniferous)

Granite and related rocks (late Silurian to early Devonian)

Sandstones and conglomerates – 'Old Red Sandstone' (late Silurian to early Devonian)

Volcanic rocks (late Silurian)

Metamorphosed sedimentary rocks – mainly Dalradian (late Precambrian)

Prominent quartzite

Limestome

Volcanic rocks and associated sills

within the Dalradian

Ancient gneisses (Proterozoic)

— · — Faults

0   5   10   15   20 Miles

0   5   10   15   20   25   30 Kilometres

1
Time and motion

Time

Argyll and the islands are richly endowed with evidence of the early human inhabitants of the area. Hunter-gatherers came to these shores some 8,000 years ago, and Bronze Age and Iron Age farmers followed hard on their heels. There is also evidence of Celtic, Pictish and Viking settlements here. For many people today, this human history seems to represent the extreme edge of time – we can go back no further. But study of the rocks extends that timescale by millions of years. For example, some of the earliest events in the geological history of Scotland are recorded in the rocks of Islay. The rocks of the Rhinns of Islay have been dated at around 1,800,000,000 (or 1.8 billion) years old. These are difficult numbers to make sense of, particularly as we tend to see all historic events through the prism of human timescales. But an appreciation of this extended timescale is an essential part of the study of geology.

Arthur Holmes, Regius Professor of Geology at Edinburgh University during the 1950s and 1960s, pioneered the accurate dating

View north over rocky shore and raised beach at Port Appin.

of rocks. His techniques delivered a degree of precision that allowed geological events to be placed in logical chronological order. This revolutionised the study of rocks and fossils and made it much easier to correlate events in one part of the country to those in another, and indeed to those in other parts of the world. Our understanding of the sequence of events that gave rise to the geology of Argyll and the islands was much enhanced by this better understanding of time.

Motion

Ours is a dynamic planet. The Earth's crust is broken into a series of individual plates that move independently of each other. The geography of the world is constantly being re-arranged as continents move and jostle their way across the globe. The unseen forces at work are powered by the energy that emanates from the Earth's core. This heat flow sets up a convection cell under the Earth's crust and, given sufficient time, these forces can move the continents many thousands of kilometres. This understanding of how the Earth works is known as 'plate tectonics'.

On their journey, continents collide with other landmasses and sometimes split apart to be separated by new oceans. The sands, muds, limestones and beds of lava that had collected on the floor of these oceans are crumpled and crushed as the continents crash into each other. There is ample evidence for this process of continental collision taking place in many parts of the world. For example, the Himalayas were formed as India split from Africa and moved northwards to collide with Asia. The resultant mountain chains are described as 'fold mountains' as the layers of soft sediments are folded, cooked and squashed in this extreme environment. As we shall see later, the Highlands of Scotland were formed in precisely this manner.

Heat flows from the Earth's core, which is a raging 6,000°C. This sets up a convection cell that moves the tectonic plates that comprise the Earth's crust. The crust is the outer layer shown in green.

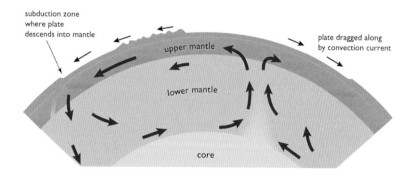

2
Ancient foundations

The oldest rocks in Argyll and the islands

We have already identified the Rhinns of Islay as the location of the oldest rocks in the area. These rocks form part of what geologists describe as 'basement' – a useful concept, as this denotes the foundation on which every other rock unit is built. The geological history of the Rhinns is complex indeed. The rocks are now much altered from their original condition when they were added to the basement in molten form many kilometres below the Earth's surface as it existed at that time. But their original character can still be determined. They were, at one time, molten magma of two slightly different compositions that cooled to become granite-like intrusions. At depth, these once molten masses were altered by heat and pressure to become

These rocks are more reminiscent of the Outer Hebrides than Islay, which is not surprising as the Rhinns of Islay and the Uists share many characteristics and a similar geological history.

Right. Laurentia, consisting of North America and fragments of Scotland, sat alongside early versions of Scandinavia, South America and Africa. This early continent, that geologists have named Rodinia, then split apart, with the creation of a seaway that widened to become an ocean – the Iapetus Ocean. It was in an area marginal to this ocean that the next important part of the story of the formation of Argyll and the islands took place. The location of the star is our best guess at Scotland's position at this time.

Below. This new seaway was also the place where the Dalradian rocks were laid down. The edge of the continent of Laurentia was divided into a series of sumps (hollows) where layers of sand, muds, and limestones accumulated under the sea. Most of the rocks described in this book were accumulated, layer upon layer, in this environment.

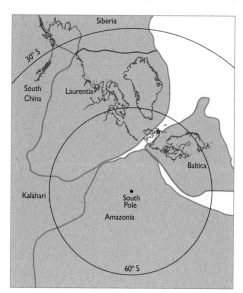

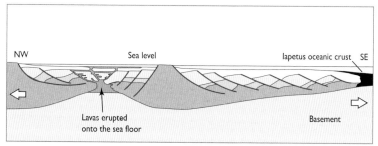

metamorphic (altered) rocks, known as gneisses (pronounced 'nices'). The rocks are banded as a result of their alteration in this extreme environment. The best place to see them is along the coast on either side of the village of Portnahaven.

World geography was very different almost 2 billion years ago. The land that was to become Scotland was part of a larger continent called Rodinia. This early slice of the Earth's crust had some unusual geographic alliances. North America sat side by side with Greenland, Scandinavia, the early forerunner of South America called Amazonia and chunks of ancient Africa. It was into this melange that the volcanic rocks of the Rhinns of Islay were injected as molten magma and then later altered by their deep burial in the early crust.

A new ocean opened

When Rodinia split apart, the newly created Iapetus Ocean played a central role in the development of the next chapter of Scotland's

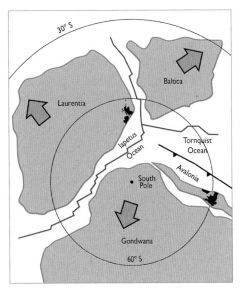

The geography of the world 550 million years ago was very different from today. Laurentia encompassed North America, Greenland and Scotland. Gondwanaland, and its off-shoots Avalonia (later England and Wales) and Baltica (which became Norway, Sweden and Finland), were separated from Scotland by the widening Iapetus Ocean.

geology. Over a period of many millions of years, this ocean increased in size to become as wide as the present-day Atlantic Ocean.

During this time, the ocean floor was piled up with layers of mud, sand, limestone and lavas that were the accumulation of debris trans-ported across the adjacent continents to the sea. The land that was to become Argyll and the islands lay close to the edge of this great ocean, and careful study of the strata within these rocks, as we see them today, indicates a variety of environments in which the layers were deposited. Sandstones alternate with limestones and slates, which are interpreted as representing different environments where the sediments accumu-lated – from deep ocean conditions, through shallow seas to layers that at one time formed part of a river delta. So geologists can establish with a considerable degree of precision a picture of changing landscapes that span around 150 million years.

Study of these rock layers has highlighted some features of inter-national importance, including some of the earliest fossils found in Scotland's geological record and layers of boulders that indicate one of the earliest glacial periods recorded anywhere in the world.

Colonsay and Uronsay

The oldest rocks to be deposited in the newly formed ocean created the islands of Colonsay and Oronsay and also the north-western part of Islay. These rocks were laid down in a shallow marine environment

The Isle of Oronsay, shown in the foreground, is separated from Colonsay by a wide tidal flat known as the Strand. Both islands share a common geological history.

close to a river delta. Younger rocks in this sequence of strata were deposited in deeper water conditions, indicating that the ocean was widening and deepening over time. The original characteristics of these rock strata have been somewhat obscured by subsequent alteration that will be described in a later section.

Lismore limestones

The rock succession that accumulated on the margins of the Iapetus Ocean consisted, for the most part, of sands, muds and boulders. These sediments were transported into the ocean by rivers that flowed across the early landscape. The terrain was barren and easily eroded as plants had not yet colonised the land. But thick limestone layers, indicative of shallow, tropical seas, are also in evidence. One of the best places to see these strata is on the island of Lismore. Almost the whole island is built from limestones. As a result, today the bedrock supports a rich and diverse ecosystem. Rocks of a similar character also

run the length of the Kintyre peninsula and across Cowal peninsula, as a thin ribbon outcrop from south-west to north-east.

The island of Lismore supports a rich ecosystem of plants and animals.

Ancient life

Colonies of bacteria, cemented in rock, are recognised as being one of the earliest signs of life in Scotland's geological record – indeed from anywhere in the world. These stromatolite fossils are found in limestones on Islay and they played an important role in the primitive ecosystem that existed in the marginal areas of the Iapetus Ocean. The colonies of cyanobacteria were the givers of life. Their gift to the early world was their ability to produce oxygen as part of their life cycle, a process we now know as photosynthesis. The atmosphere in those early times was fairly toxic to life as we know it today. The colonisation of the world's oceans by stromatolites helped to make the planet a more hospitable place to the new forms of oxygen-breathing life that were to follow. *Homo sapiens* and all the other species of air-breathing animals on Planet Earth owe a great deal to these simple bacteria.

Stromatolites are survivors. They still exist today and are one of the very few forms of life that have changed little through many hundreds of millions of years of geological time. Shark Bay in Australia has thriving colonies of stromatolites – a throwback to the planet's earliest flirtation with life on Earth.

Snowball Earth – a worldwide glaciation

Rocks found around Port Askaig on Islay provide evidence for a 635-million-year-old glaciation. The layers of assorted boulders and mud found there are around 750m thick and have been interpreted as 'fossilised' glacial deposits. We are familiar with glaciations that occurred during more recent times, specifically the events that shaped the landscapes as we see them today. But glacial events have now been identified at other times in the geological past. These boulder-rich layers, known as tillites, also occur on the Garvellach Isles and Jura and have also been correlated with similar deposits in Perthshire, notably what has become known as the Schiehallion boulder bed.

The tillite layers in Scotland, and many similar ones of the same age found on every continent across the globe, provide evidence for an idea that the whole planet was affected by an ice age over 600 million years ago. This hypothesis suggests that frozen wastes extended as far south as the Equator, so the Earth was like a giant snowball, hence the title Snowball Earth. After these short, sharp glacial episodes,

Stromatolites from Shark Bay Australia. They may look fairly unimpressive structures, but they helped to change the way life on Earth developed.

These 'boulder beds', known as 'tillites', at Port Askaig on Islay have been interpreted as the deposits formed as a result of glaciation. They are made up of assorted pebbles and boulders that were laid down in shallow seas between the main glacial phases when the climate was milder.

global temperatures are thought to have soared during later 'greenhouse' phases. These rapid and widespread changes in global temperatures demonstrate that climate change isn't a recent phenomenon. It has occurred throughout geological history, and future global temperatures will also be subject to considerable natural variations.

A thick blanket of sand

Jura has one of the most distinctive profiles of any Scottish island. The Paps of Jura can be seen for miles around. The form of these hills is related to the way they were moulded during the Ice Age in more recent geological times, but the material they were originally created from is also of interest. After temperatures stabilised from their wide fluctuations from cold to much warmer conditions, a thick blanket of sand was laid down across a wide beach area that was subjected to the ebb and flow of the tides. The fact that these sand deposits are so thick indicates that the floor of the ocean was sinking at that time, but the shallow water depth was maintained by the rapid and copious input of sand to this wide tidal shelf area. Thick coarser layers made up of cobbles and small boulders (conglomerate) also form part of this geological sequence, reinforcing the interpretation that these rocks represent a high-energy coastal environment. After the closure of the Iapetus Ocean, in the extreme conditions generated as continents collided, these pure sandstones were altered to quartzites that are as hard as flint.

The Paps of Jura are a distinctive and visible landmark of the area. They are made from sandstones that were subsequently altered to durable quartzites. Their current distinctive shape was sculpted during more recent times when the area was covered by snow and ice.

Underwater eruptions

As the Iapetus Ocean continued to widen, lavas were erupted onto the ocean floor from a series of underwater volcanoes. The lavas cooled immediately as they were belched out into the cold seawater. They formed pillow structures that are characteristic of this style of eruption. Rapid cooling led, as similar eruptions do today, to the formation of a thin glassy crust around each pillow. These pillow lavas, found on the coast of the Tayvallich peninsula, have been dated as 600 million years old.

Above. These pillow lavas on the Tayvallich coast, are stacked into a pile, which is characteristic of this style of eruption.

Right. This image shows lavas being erupted underwater today. The molten rock forms into pillow-shaped blobs that are stacked one on top of the next.

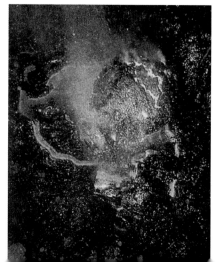

The Iapetus Ocean begins to close

Around 500 million years ago, the Iapetus Ocean began to close. This re-arrangement of the world's continents and oceans was driven by the on-going ferment in the Earth's mantle below. Layer upon layer of sands, muds, limestone and lava had built up on the sea floor since the Iapetus Ocean became established around 750 million years ago. As continents converged, the process of building what we now recognise as the Scottish Highlands began.

This Celtic cross, known as the Kildalton High Cross, was fashioned from ancient lavas of the type described on the facing page.

Granites galore

The pressure generated by colliding continents was intense. These high pressures, coupled with the heat of burial deep within the Earth's crust, caused the lower parts of the collision zone to melt, forming copious quantities of molten magma of granite composition. The reservoir of molten rock that became the Etive granite in the north of Argyll was

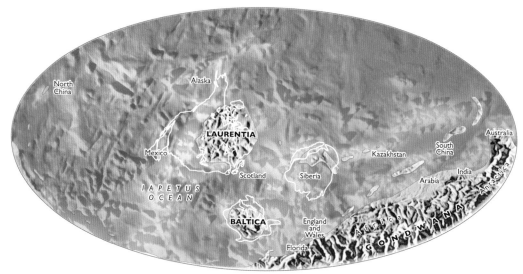

This was the arrangement of land and sea at the time the Iapetus Ocean began to close. 'Scotland' was a little piece of North America, while England and Wales lay close to the South Pole. As the ocean closed, the continents converged.

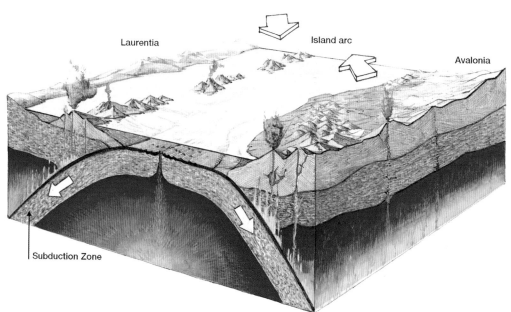

As the ocean closed, the leading edge of one of the tectonic plates dipped under 'Scotland' to create a subduction zone. As the continents moved ever closer, the ocean-floor sediments were squeezed and altered by the increased heat and pressure. Over time, a mountain range as high as the Himalayas was formed. All the rock sequences described above were affected. From the very oldest rocks laid down in this ocean (the rocks of Colonsay), through the Lismore limestones, stromatolites, glacial deposits of Snowball Earth, and the Tayvallich ocean floor lavas, all these rocks were folded and intensely squeezed to the point that their original character became difficult to discern. They were metamorphosed, fundamentally changed by this process of elevated heat and pressure.

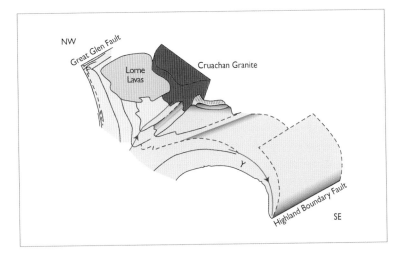

The ocean-floor sediments were bent and buckled as a result of this major continental re-arrangement. The heat was so intense that the lower layers of this chaotic pile melted to form granite and related rocks. These molten magmas then flowed back to a higher position in the Earth's crust as they were less dense than surrounding rocks.

trapped at depth and crystallised there. Elsewhere, however, this molten rock was able to find its way back to the surface and spilled out across the surface as lava. These lava flows were relatively fluid and were able to run far and wide across the ancient land surface as it existed around 425 million years ago. These long-solidified flows of rock are recognised as building the foundations of the area around Oban and the surrounding stepped hillside, collectively known as the Lorne Plateau.

The stunning mountains at the head of Loch Etive were carved from the granites that were generated by colliding continents.

3
After the storm

Devonian deserts

Thick layers of conglomerate, consisting of a variety of rock types, were deposited after being transported downslope by streams and rivers. The boulders are now rounded, because the angular corners were knocked off as they were carried along down the mountainsides. These rocks are dated at around 420 million years old.

Around 410 million years ago, the land that was to become Scotland was now part of an arid continent that lay just south of the Equator. The high mountains, created by continental collision, were rapidly eroded by the rivers and streams that ran down the parched slopes. There was little vegetation in the upland areas to slow down this aggressive process. Dramatic flash floods tore down the fabric of the mountains as streams carried a colossal load of sand, pebbles and boulders to the lower ground. The evidence for these events is preserved in deposits of conglomerate that are found in a variety of places, but perhaps best represented on the island of Kerrera. Great

boulders that once formed part of the towering mountain chain are now embedded in deposits that accumulated in the valley floor.

    Rocks of a similar age are also found around Oban, exposed in some road cuttings and also along the very tip of Mull of Kintyre from Campbeltown southwards.

Skerry Fell Fad near Machrihanish is a dome of trachyte lava that erupted some 340 million years ago during Carboniferous times.

Volcanoes and coal swamps

The Carboniferous Period started with a bang in this area! In early Carboniferous times, around 340 million years ago, basalts erupted to form terraced hillsides, as well as stickier lavas of a more silica-rich composition, called trachytes, which formed prominent landscape knolls.

Making coal

Around 330 to 300 million years ago, the continent of which Scotland was a part sat astride the Equator. By this time, plants were a vibrant part of the burgeoning ecosystem, and thick forests covered the land. The sea level oscillated quite markedly during this time as a result of

This is what the landscape would have looked like when coal-forming forests covered the land. Tree-ferns, horsetails and lepidodendrons formed a dense canopy of vegetation. Dragonflies grew to massive sizes with a wingspan of over 0.5m because oxygen levels in the atmosphere were much higher than they are today.

carly and on-going changes in climate. So the tropical rainforests that covered much of 'Scotland' during these far-off times were regularly inundated by the sea. As a result, the submerged vegetation died back and was covered by layers of sand and mud dumped by the advancing deltas. When sea levels fell, the forest re-established and another cycle was initiated. As the sands, mud and vegetation piled up layer upon layer, the vegetation was compressed and hardened to form carbon-rich seams – coal.

These layers lay buried for hundreds of millions of years, until they were unearthed during the Industrial Revolution. It seems slightly incongruous, but amid the scenic grandeur of the Scottish Highlands, a small coalfield was discovered near Machrihanish. The coal seams are not exposed at the surface. They are covered by beach deposits and blown sand. This tiny fragment of the coal-bearing strata that at one time covered much of central Scotland was preserved because it is bounded by geological faults. A block of coal-bearing strata dropped down to the same level as the surrounding harder rocks and was there-fore protected from subsequent erosion. The outcrops here, and similar rocks at the Pass of Brander, add vital evidence to support the

fact that tropical forests cloaked the southern half of the country from east to west.

Work in progress at the Machrihanish coal mine.

Coal seams up to 3 metres thick were worked from the end of the eighteenth century onwards, and the Machrihanish mine was only closed in 1967 after a serious fire.

Desert sands

During Permian times around 300 million years ago, as 'Scotland' moved north, driven by the movement of the continents, the area that was to become Argyll and the islands lay at a similar latitude to the Sahara desert. It is no surprise then that the land was, once again, part of an arid desert where sand dunes were the dominant landscape feature. Few plants or animals could survive in this ferocious environment where withering desert storms scarified the early landscape. Little evidence survives of this extreme environment today, but some exposures of brick-red sandstone that date back to these times can still be seen along the Atlantic coast of the Mull of Kintyre.

Mind the gap!

Red sandstones of Permian age form prominent roadside features between Bellochantuy and Tayinloan.

A period of around 187 million years elapsed before the next geological event left a permanent mark on the landscape. These periods of hiatus, where substantial gaps in the record of the rocks occur, are not unusual. Significant geological events must have been happening around Argyll and the islands. Perhaps most noteworthy was the substantial rise in sea level that took place during Jurassic and Cretaceous times. During these periods, much of the land that was to become Scotland was submerged under the waves as global sea levels rose to new and unprecedented heights. Thick carpets of sands, muds and limestones were laid down in the area we now recognise as Skye. This area was also home to a range of plant-eating and meat-eating dinosaurs whose fossil remains and footprints have been unearthed from the rocks in that area. There may have also been similar events in Argyll and the islands, but if the evidence did exist, it is now long gone, wiped away by subsequent erosion.

Vertical sheets of magma

When the geological record resumes in this area, it is in the form of vertical sheets of molten rock, called dykes, that cut through the older rocks of the area. These pulses of magma originated from momentous events that took place a few kilometres to the north-west – the eruption of the Mull volcano. This was a period when the continents underwent yet another major re-arrangement: the North Atlantic Ocean was beginning to emerge as continental Europe moved away from North America. Heat flowed from the upper mantle of the Earth as a result of this separation and a line of volcanoes became active down the north-western seaboard of Scotland. The Mull volcano propelled thin vertical sheets of molten rock along a south-easterly and easterly linear trajectory across the Scottish mainland. They are represented on the geological map on page 10 of this book by a series of almost parallel lines that cut across the bedrock.

These linear features are 60-million-year-old dykes from the Mull volcano that stand proud from the beach on the west coast of Jura.

4
The Ice Age

The Ice Age was the final natural event that shaped the landscape we see today. Over the last 2.6 million years, the climate has fluctuated between temperatures similar to that of today to prolonged cold periods where thick layers of ice and snow were a permanent feature throughout the year. During these cold periods, northern latitudes were submerged under a blanket of snow and ice. Sea levels fell world-wide as much of the water on the planet was locked in glacial ice and snow. Short spells of warmer conditions known as inter-glacials punctuated these prolonged cold periods.

For an explanation for these dramatic temperature fluctuations, we need to look to space. The Earth has an eccentric orbit around the Sun, varying from a perfectly circular passage around our home star to something much more elliptical. The period of transition from one orbit to the next takes around 100,000 years. When the planet is furthest from the Sun, less sunlight falls on the Earth's surface and this is sufficient to tip the climate into a cold phase. Variations in the Earth's tilt and wobble as it spins on its axis also affect the amount of solar

This is Antarctica today, but it's also how Scotland would have looked during the last advance of the ice. Even the highest peaks in Argyll and the islands would have been submerged under ice and snow. These advances and retreats of the ice had a significant effect in shaping the landscape, with mountain and glen carved and moulded by the passage of the ice.

radiation that reaches ground level. During the last 2.6 million years, there have been many such transitions from warm to cold and many more are expected in the future. We currently bask in the warmth of an inter-glacial period, but at some point into the future, perhaps as soon as in the next 50,000 years, the glaciers will return and, once again, the country will be caught in the icy thrall of the next glacial advance.

The last glaciation reached its peak around 22,000 years ago when vast sheets of ice flowed westwards across the area from the highest ground towards the sea. As the global temperatures rose slightly, the ice cover thinned and the glaciers became confined to the glens and sea lochs that had previously been carved out by the ice. Where the glaciers met the sea, the ice would have broken into large chunks forming icebergs, as happens in Greenland and Alaska today.

The Paps of Jura emerged fairly early from the blanket of ice, and the bedrock of quartzite was then subjected to extreme freezing conditions. These conditions had a shattering effect on the rock, and as pieces were broken off the upper slopes they tumbled down the hill to accumulate as an extensive scree slope. This jumble of rock was added to, even after the ice had finally disappeared but the climate remained harsh.

Temperatures rose suddenly around 14,700 years ago and pioneer plants became established across the post-glacial landscapes. But this respite from the harsh conditions was temporary. Shortly afterwards, the climate began to cool, and ice once again accumulated on the higher ground. Evidence for this climate cameo is provided by pollen and spores preserved in a silted-up loch near Oban.

An impressive build-up of frost-shattered blocks of quartzite accumulates on the side of the Paps of Jura.

5
After the ice

A new beginning

Below. The European brown bear, which used to inhabit this area.

Below right. The European beaver used to be a common sight around Argyll and the islands but our activity caused this charismatic animal to become extinct across the country. However, 'what goes around comes around', and this species has now been re-introduced to the Knapdale area and the population is already expanding. The reappearance of the beaver has not been universally acclaimed, particularly by local farmers and fishermen.

The climate started to warm abruptly again around 11,500 years ago. The ice melted and, over time, a burgeoning ecosystem became established. Plant species such as heather, juniper and a range of grasses were among the first to gain an early toehold on the moonscape of jumbled deposits dumped by the ice. We can piece together the ecological succession of events because pollen grains, early insects and other microfossils have been found in lake sediments and peat bogs across the area. Many mammals that were common in those early days are now extinct in Scotland – animals such as the European brown bear, lynx and wolf. This was the world that people first encountered when they settled the area around 8,000 years ago. Those early immigrants modified the ecosystem to suit their way of life and, as a consequence, the substantial forest that covered the area was gradually chopped down to make way for an increasingly pastoral way of life. Widespread cereal and flax cultivation is recorded in the pollen record from around this time.

Changes at the coast

Today, we fret about changes of a few centimetres in the sea level, which is perhaps understandable if your house happens to be close to high water mark or the edge of a sea cliff. But, at the end of the Ice Age, adjustments in sea level were massive in comparison. Huge quantities of water had been added to the world's oceans as temperatures rose and ice sheets melted all across the globe. But the land was also on the move – upwards. During the Ice Age, the whole landmass was weighed down by a thick layer of ice for millennia, so when the ice sheets melted, the land 'bounced' back, albeit ever so slowly. This process is known as 'isostatic readjustment' and can raise landscapes by tens of metres. It was the interplay between the higher global sea levels and landmasses recovering their former position after their ice burden was lifted that determined the position of the coastline. The location of the coastline changed on a number of occasions as each of these variables became more or less significant.

This is a reconstruction of how the landscape might have looked shortly after the ice melted. It would have been a hostile environment for the early immigrants, but these early settlers were resourceful and created an environment that allowed a sustainable way of life to be established.

Right. This simple graph indicates the changes in sea level in the Knapdale area from 17,000 years ago to the present day. As the ice melted, sea levels rose massively to almost 40 metres above their present levels. Many features such as sea arches, cliffs cut by the sea and beach deposits now occur tens of metres inland above the high-water mark. The rebound of the land after the ice burden was lifted then caught up and the rise in sea level relative to the land was substantially reduced.

Below. This is a sea arch created by coastal erosion around 12,000 years ago when sea levels were much higher than today. Locally, it is called Clach Tholl in Gaelic, meaning 'hole in the rock'.

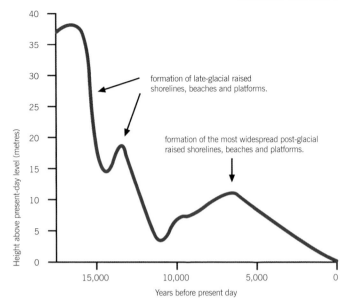

formation of late-glacial raised shorelines, beaches and platforms.

formation of the most widespread post-glacial raised shorelines, beaches and platforms.

Above. The retreat of the sea lapping at the foot of the highest cliffs on the right of this picture is clearly evident. Sea level fell in a series of steps, and again this is reflected in the landscape, as a second sea cliff has been cut in the centre of this view. Finally, sea level fell again relative to the land and exposed a wide area, described as a raised beach, now grazed by sheep. This is on the south coast of Kerrera.

Left. The Dog Stone near Oban provides yet more evidence of sea levels that were once much higher than today. This sea stack was fashioned from conglomerate outcropping at the coast and then left high and dry when the sea level retreated relative to the land. According to legend, the giant Fingal tethered his dog Bran here while he went hunting.

6
Landscapes today

Machair landscapes

Below. Machair flowers.

Opposite top. Machair in full bloom.

Opposite bottom. Daisies and silverweed are commonly found in the machair sward.

Machair is the strip of fertile low-lying land that lies between the sea and harder rocks inland. The soil is made from broken shells, smashed by the sea offshore and in the inter-tidal area, and then blown inland by the prevailing wind. This fertile land has been farmed by local crofters in a low-intensity manner for many generations. The land supports a few head of cattle and sheep, with oats and potatoes being the favoured crops. During the summer, this strip of land is ablaze with many species of flowering plant. Orchids, silverweed and clovers create a dazzling tableau of colour.

This aerial view of the Kyles of Bute demonstrates the tranquillity of this special place, which is a haven for sailors and walkers to enjoy the peace and stimulation of being close to nature.

Aesthetic landscapes

The Kyles of Bute is a narrow sea channel that separates the north end of the Isle of Bute from the Cowal peninsula and is one of the scenic jewels of Scotland, designated as a National Scenic Area (NSA). Part of the site description reads: 'The enclosing hillsides coming down to the sea are clothed in extensive areas of mature, mixed woodland. These possess a verdant and luxuriant canopy, often reaching the water's edge. Individual mature trees spread their boughs across small pastures, glades and clearings.'

The islands of Scarba, Lunga and the Garvellachs are also designated as a National Scenic Area. The official NSA description of this collection of islands is also lyrical and encourages the visitor to make the journey 'Set between the large mass of Jura and the Argyll mainland, this arc of islands presents a remote and isolated aspect. They are far removed from urban centres, accessible only when sailing conditions are favourable. Today there is no permanent habitation on any of the islands, although the remains of past settlement show that it has not always been so. While there is still some grazing by sheep and cattle, the islands nowadays have a wild and undeveloped appearance with an absence of overt human influence that belies their history.'

Between Scarba and Jura lies the fearsome Gulf of Corryvreckan. On a spring tide, water is drawn through the Sound of Jura and a bottleneck is created. This is further compounded by the presence of a pillar of basalt rock that sticks up from the sea floor at a depth of 70 metres, the top of which is almost 30 metres below the surface of the sea. The water funnelling through the straits between the islands creates a whirlpool that has bedevilled sailors for generations. Many lives have been lost in these turbulent waters. In earlier times, in the absence of a scientific explanation, mythology and legend filled the gap in understanding this natural phenomenon. Scottish mythology tells us that the hag goddess of winter, Cailleach Bheur, used the gulf to wash her great plaid, and this ushered in the turn of the seasons from autumn to winter. However, the presence of an underwater basalt pillar seems a more prosaic and likely explanation.

Gulf of Corryvreckan

George Orwell wrote *Nineteen Eighty-Four* at Barnhill farmhouse on Jura.

This windfarm near Campbeltown dominates the landscape, but it undoubtedly returns a benefit to the country's economy.

Literary landscapes

Barnhill farmhouse at the northern end of Jura overlooks the Gulf of Corryvreckan. It was to this place that George Orwell retreated to write *Nineteen Eighty-Four*, his most significant work. Orwell's diary entry from 17 August 1947 described a boating trip and an encounter with Corryvreckan thus: 'On return journey, ran into a whirlpool and we were all nearly drowned. Engine torn off by the sea and went to the bottom. Just managed to keep the boat steady with the oars and then we ran into calmer waters . . . It appears that this was the very worst time, and we should time it so as to pass Corryvreckan in slack water. The boat is alright. The only serious loss, the engine and 12 blankets.'

Working landscapes

Renewable energy is now big business. In the years immediately after the Second World War, hydroelectric dams were constructed to fill many of the Highland glens with water to generate vital energy for the recovering Scottish economy. Now engineers have turned their attention to harvesting energy from another plentiful and endlessly renewable resource – the wind. As with hydro schemes, this has had an impact on the landscape. Many schemes have created controversy

as well as electricity, but it is undeniably a boon for some island communities where wind turbines are run as a community enterprise.

The Cruachan Power Station, located just to the east of Oban, is another example of a working landscape. This pumped storage scheme was installed in 1965 and involved creating tunnels deep into the mountainside and the hollowing out of a chamber large enough to accommodate the turbine hall. When required, water cascades from the small dammed reservoir loch (see photo) to spin the four massive turbines in the turbine hall situated immediately below the loch. The power station is brought into service when it is anticipated there will be a power surge on the National Grid, such as at full time after a live televised football match or at the end of a popular TV programme when many will be switching on their kettles. From the turbine hall, the water then runs into the adjacent Loch Awe. At night, when electricity is much cheaper and the demand substantially lower, the turbines are reversed and the water is pumped back up into the reservoir loch from Loch Awe. The station then awaits the next call to meet a spike in electricity demand.

Aerial view of the Cruachan pumped storage scheme, which shows the dammed reservoir loch. All of the inner workings of the power station are concealed in the heart of Ben Cruachan. The mountain is made from Etive granite, prominent in this area of Argyll.

All the ingredients of a fine dram

Whisky is made in all parts of Scotland, but there are only two places where whisky making completely dominates lives and landscapes – Speyside and Islay. On Islay, there are eight distilleries producing whisky of the highest quality. All these distilleries take water from burns, lochs and lochans across the island that are broadly similar in character and chemistry. What distinguishes these restorative liquors one from another is the extent to which the barley is infused with peat smoke during the malting process. Some of the darkest and most medicinal single malts come from this island – Laphroaig, Lagavulin and Ardbeg. They are not to everyone's taste, but they assisted in no small measure with the writing of this book.

Some of the peats cut on Islay are used for burning in the hearth, but some are used for flavouring the malt whiskies for which the island is rightly world-renowned.

7
Places to visit

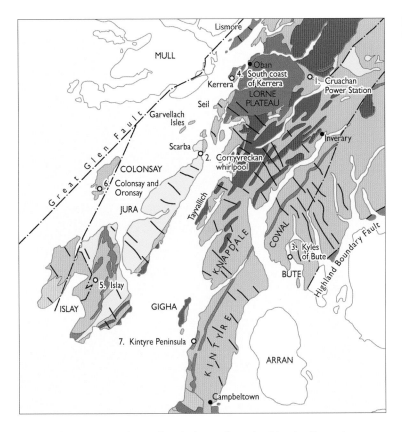

Map showing the locations of places to visit.

Many places to visit have already been described in the foregoing text, so this short chapter simply highlights some of the best places to see the geology and landscapes of the area at first hand. There are many OS Landranger sheets that cover the area including 49, 55, 60, 61, 62 and 68.

Further details of places of natural heritage interest (geological, biological and aesthetic) can be found on the SNHi portal at www.snh. gov.uk/snhi.

1. **The Hollow Mountain:** the Cruachan Power Station is located some 20 kilometres east of Oban and is well geared up to receive visitors. A description of how the place operates is provided on page 41.

2. **Corryvreckan:** the whirlpool between Jura and Scarba is described on page 38.

3. The Kyles of Bute: this tranquil place is again described on page 38.

Right. South coast of Kerrera

Below. Famous distillery on Islay.

4. South coast of Kerrera: the raised beaches of Kerrera are described on page 35.

5. Islay: a visit to Islay would not be complete without a tour of at least some of the many distilleries on the island. Assuming that task has been completed, the footpaths around Portnahaven on the Rhinns of Islay take the visitor to the oldest rocks on the island. There is also adequate parking at this location.

Above. Strand linking Colonsay and Oronsay.

Left. Roadside exposure of Permian sandstone.

6. Colonsay and Oronsay: these islands, linked by a tidal causeway, are off the beaten track, but are an interesting geological destination.

7. Kintyre peninsula: desert sandstones of Permian age and adjacent raised beach features make the western coast of Kintyre an interesting stopping point. These features are immediately adjacent to the road, with roadside parking available.

Acknowledgements and picture credits

Thanks are due to Professor Stuart Monro OBE FRSE and Moira McKirdy MBE for their comments and suggestions on the various drafts of this book. I also thank Debs Warner, Mairi Sutherland, Andrew Simmons and Hugh Andrew from Birlinn for their support and direction. Mark Blackadder's book design is up to his usual high standard. Scottish Natural Heritage, in association with the British Geological Survey, published the *Landscape Fashioned by Geology* series that was the precursor to the new *Landscapes in Stone* titles. I thank them both for their permission to use some of the original artwork and photography in this book. David Stephenson and Jon Merritt wrote the original text for *Argyll and the Islands – A Landscape Fashioned by Geology*, which influenced aspects of this book. Another very useful source was *A Guide to the Geology of Islay* published by Ringwood Publishing. I dedicate this book to Hugh Andrew, founder and managing director of Birlinn. He has entrusted me with the task of telling Scotland's amazing geological story in a manner that will appeal to as wide an audience as possible. I am grateful to Hugh for giving me this unique opportunity.

Picture Credits

Title page Andy Sutton/Alamy Stock Photo; 6 Dennis Hardley/Alamy Stock Photo; 10 drawn by Jim Lewis; 11 Lorne Gill/SNH; 12 drawn by Jim Lewis; 13 David Bell/Ecos; 14 (upper and lower) drawn by Jim Lewis; 15 drawn by Jim Lewis; 16 Pat & Angus Macdonald/Aerographica/SNH; 17 Pat & Angus Macdonald/Aerographica/SNH; 18 (lower) Lorne Gill/SNH; 19 Fergus MacTaggart; 20 (upper) Lorne Gill/SNH; 20 (lower) Katia & Maurice Krafft; 21 Jamie Pharr; 22 (upper and lower) drawn by Robert Nelmes; 23 (upper) drawn by Jim Lewis (lower) John A. Cameron; 24 Lorne Gill/SNH; 25 Lorne Gill/SNH; 26 Clare Hewitt; 27 © Scottish Mining Museum; 28 Lorne Gill/SNH; 29 Iain Thornber; 30 John Gordon; 31 Iain Thornber; 32 (left) BMJ (right) Arterra Picture Library/Alamy Stock Photo; 33 Craig Ellery; 34 (lower) Lorne Gill/SNH; 35 (upper and lower) Lorne Gill/SNH; 36 Lorne Gill/SNH; 37 (upper) Martin Fowler (lower) Lorne Gill/SNH; 38 Pat & Angus Macdonald/Aerographica; 39 Pat & Angus Macdonald/Aerographica/SNH; 40 (upper) Mary Evans/Everett Collection (lower) Lorne Gill/SNH; 41 Dennis Hardley/Alamy Stock Photo; 42 Lorne Gill/SNH; 43 drawn by Jim Lewis; 44–45 Dave Head; 46 (upper) Lorne Gill/SNH (lower) Lukassek; 47 (upper) Richard Clarkson/Alamy Stock Photo (lower) Alan McKirdy.